UNDERFOOT

David M. Schwartz *is an award-winning author of children's books, on a wide variety of topics, loved by children around the world.* Dwight Kuhn's *scientific expertise and artful eye work together with the camera to capture the awesome wonder of the natural world.*

For a free color catalog describing Gareth Stevens Publishing's list of high-quality books and multimedia programs, call 1-800-542-2595 (USA) or 1-800-461-9120 (Canada). Gareth Stevens Publishing's Fax: (414) 225-0377.

Library of Congress Cataloging-in-Publication Data

Schwartz, David M.
    Underfoot / by David M. Schwartz; photographs by Dwight Kuhn.
        p. cm. — (Look once, look again)
    Includes bibliographical references (p. 23) and index.
    Summary: Explores the world of ants, centipedes, earthworms, star-nosed moles, and other animals living in or on the ground.
    ISBN 0-8368-2246-3 (lib. bdg.)
    1. Soil animals—Juvenile literature. [1. Soil animals.]
I. Kuhn, Dwight, ill. II. Title. III. Series: Schwartz, David M.
Look once, look again.
QL110.S39   1998
591.75'7—dc21   98-6307

This North American edition first published in 1999 by
Gareth Stevens Publishing
1555 North RiverCenter Drive, Suite 201
Milwaukee, Wisconsin 53212 USA

First published in the United States in 1997 by Creative Teaching Press, Inc., P.O. Box 6017, Cypress, California, 90630-0017.

Text © 1997 by David M. Schwartz; photographs © 1997 by Dwight Kuhn. Additional end matter © 1999 by Gareth Stevens, Inc.

Printed in the United States of America

1 2 3 4 5 6 7 8 9 03 02 01 00 99

UNDERFOOT

by David M. Schwartz
photographs by Dwight Kuhn

A SPRINGBOARDS INTO
SCIENCE
SERIES

Gareth Stevens Publishing
MILWAUKEE

These are not little fingers. They are the fleshy feelers of a very slimy animal, and they are looking right at you!

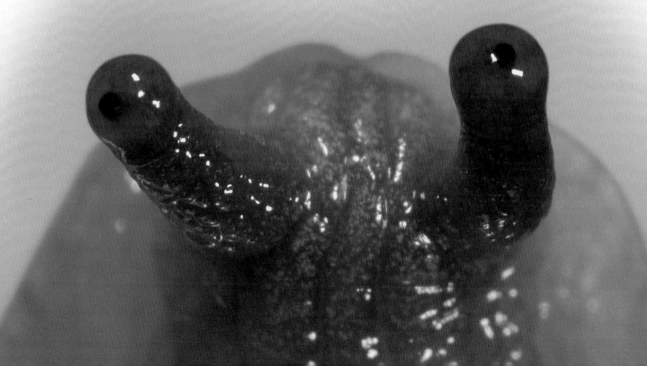

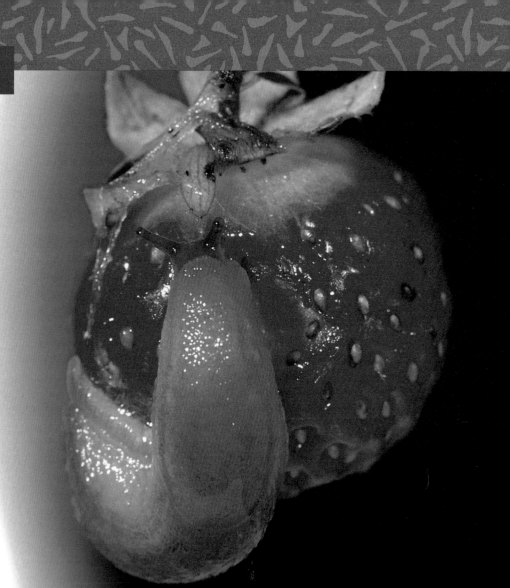

Slugs are snail-like animals without shells. Their tiny eyes are on the ends of their feelers, or tentacles.

Slugs glide along on a layer of slime, looking for good things to eat. They like strawberries and many other tasty fruits. If you like strawberries, you should eat them before the slugs do!

No wonder its face is dirty! This animal spends most of its time "worming" its way through the earth.

Earthworms have no eyes, no nose, and no ears.
But they have a mouth. They eat dead
plants and animals in the ground.
Their waste makes the soil very
rich, so more things can grow.

They also loosen the soil
so air can get in, and
water can drain away.
Earthworms do
very important
jobs in the garden.

Is it a star, or is it a nose?
Or is it a starry nose?

It is a star-nosed mole!
This mole does not use
its nose to sniff for food
but to feel for food.
The twenty-two fleshy
tentacles on the mole's
nose help it find juicy
worms and yummy bugs.

This is not the shining armor of a knight.
It is the armor of a bug that wants its enemies to "bug off."

11

A sow bug has small, hard plates that help protect it from predators. Sow bugs live in dark places. In the light, they quickly scurry away, looking for a dark place to hide.

Legs, legs, and more legs. Some people think this creepy crawly has a hundred legs.

It is a centipede. The word *centipede* means "one hundred legs."
Most centipedes have thirty to forty-six legs, not one hundred.
Watch out for the front legs. They are called "poison claws."
You can probably guess why. The centipede uses its poison
claws to kill its prey.

You can find these shiny brown ridges on something that grows in forests and other damp places. You may have to get on your hands and knees to see them.

They are gills of a mushroom. Gills make thousands of tiny spores. The spores grow into more mushrooms.

Never eat a wild mushroom because some mushrooms are poisonous.

It is small, it is black, it
is f-ANT-astic!

17

This ant looks like it is attacking the other insect, but it is not. The other insect is a treehopper. The treehopper makes a sweet liquid that ants like to drink. The ant is described as "milking" the treehopper.

Look closely. Can you name these plants and animals?

A. Slug

B. Earthworm

C. Star-nosed mole

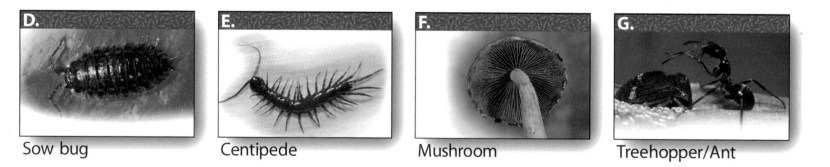

D. Sow bug

E. Centipede

F. Mushroom

G. Treehopper/Ant

How many were you able to identify correctly?

armor: a protective covering, like the hard shell of some animals.

feeler: a slender, flexible part of an animal that is used for touching or feeling.

gill: one of the spore-producing parts of a mushroom found on the underside of the mushroom's cap.

mole: a small, burrowing animal with very small eyes and short, silky fur.

plate: a hard, thin part of the armor of some animals, like the sow bug.

predator: an animal that hunts other animals for food.

prey: an animal hunted by another animal for food.

ridge: a narrow, raised strip.

slug: a small, soft-bodied land animal that is like a snail without a shell.

sow bug: a type of small crustacean that lives in damp and dark places. Sow bugs are also called pill bugs, because they can roll themselves into a tight ball, like a pill.

spore: a single plant or animal cell that is able to grow into a new plant or animal. Mushrooms reproduce by means of spores.

tentacles: the narrow, flexible parts of certain animals used for feeling, grasping, and moving.

waste: natural material that is left after food has been eaten and digested by an animal.

worming: moving like a worm.

ACTIVITIES

Dig It!
What do you notice about the star-nosed mole that makes it a good digger? Make a list of other animals that live in burrows in the ground. Do all of these animals dig their own holes, like the star-nosed mole? Find books in the library and search the Internet to learn more about the different kinds of animals that live underground.

Spore Art
You can clearly see the spores of a gilled mushroom after making a print on a dark piece of construction paper. Remove a mushroom's stem, and place the flat side of its cap onto the paper. Put a small bowl over the mushroom cap to keep it out of drafts. After a few hours, remove the bowl. Can you see the pattern of spores on the paper? Do not eat the mushroom, and wash your hands after handling it.

How Now, Sow Bug?
Sow bugs like to live in dark, damp places. Lift up stones, bricks, or pieces of wood to find them. Then do an experiment to learn more about them. Put a dry paper towel on one-half of a shallow box and a wet paper towel on the other half. Carefully put some sow bugs into the middle. Watch them for about thirty minutes. What do you observe about the sow bugs' behavior? Return them to the wild when done.

Make a Wormery
To make a wormery, ask an adult to cut the top off a large, clear, plastic container. Alternate layers of soil and sand in the container. Put a few worms on top and cover them with dead leaves. Put the wormery in a dark cupboard. What do you see happening from day to day? Return the worms to the wild after your observations.

More Books to Read

Ants: A Great Community. Secrets of the Animal World (series). Andrew Llamas (Gareth Stevens)
Centipedes. The New Creepy Crawly Collection (series). Graham Coleman (Gareth Stevens)
Earthworms: Underground Farmers. Patricia Lauber (Henry Holt)
Moles: Champion Excavators. Secrets of the Animal World (series). Eulalia García (Gareth Stevens)
Mushroom. Barrie Watts (Franklin Watts)
Snails and Slugs. Chris Henwood (Franklin Watts)

Videos

Ants and Worms. (Agency for Instructional Technology)
Life on the Forest Floor. (Phoenix/BFA Films & Video)
The Snail. (Barr Films)
Underground Animals. (Wood Knapp Video)

Web Sites

www.icon.portland.or.us/education/vose/kidopedia/ants.html
www.cybercom.net/users/dhe/Meadow/howwormswork.html

Some web sites stay current longer than others. For further web sites, use your search engines to locate the following topics: *ants, centipedes, moles, mushrooms, slugs, sow bugs,* and *worms.*

INDEX